OLD MEN AND C

By the same author:

Collected Poems 1987 (Oxford Poets)
Selected Poems 1990 (Oxford Poets)
Under the Circumstances, 1991 (Oxford Poets)

OLD MEN AND COMETS

D. J. Enright

Oxford   New York
OXFORD UNIVERSITY PRESS
1993

Oxford University Press, Walton Street, Oxford OX2 6DP
Oxford New York Toronto
Delhi Bombay Calcutta Madras Karachi
Kuala Lumpur Singapore Hong Kong Tokyo
Nairobi Dar es Salaam Cape Town
Melbourne Auckland Madrid
and associated companies in
Berlin Ibadan

Oxford is a trade mark of Oxford University Press

© D.J. Enright 1993

First published in Oxford Poets
as an Oxford University Press paperback 1993

All rights reserved. No part of this publication may be reproduced,
stored in a retrieval system, or transmitted, in any form or by any means,
without the prior permission in writing of Oxford University Press.
Within the UK, exceptions are allowed in respect of any fair dealing for the
purpose of research or private study, or criticism or review, as permitted
under the Copyright, Designs and Patents Act, 1988, or in the case of
reprographic reproduction in accordance with the terms of the licences
issued by the Copyright Licensing Agency. Enquiries concerning
reproduction outside these terms and in other countries should be
sent to the Rights Department, Oxford University Press,
at the address above.

This book is sold subject to the condition that it shall not, by way
of trade or otherwise, be lent, re-sold, hired out or otherwise circulated
without the publisher's prior consent in any form of binding or cover
other than that in which it is published and without a similar condition
including this condition being imposed on the subsequent purchaser.

British Library Cataloguing in Publication Data
Data available

Library of Congress Cataloging in Publication Data
Enright, D. J. (Dennis Joseph), 1920–
Old men and comets / D. J. Enright.
p.   cm. — (Oxford poets)
I. Title. II. Series.
PR6009.N6053 1993 821'.914—dc20 93-20417
ISBN 0-19-283176-3

1 3 5 7 9 10 8 6 4 2

Typeset by J&L Composition Ltd, Filey, North Yorkshire
Printed in Hong Kong

Acknowledgements

Acknowledgements are due to the editors of the following periodicals, in which some of these poems first appeared: *The Applegarth Review*, *The Independent on Sunday*, *London Magazine*, *London Review of Books*, *P.N. Review*, and *The Times Literary Supplement*.

Contents

Clichés 11
Coming true 11
In the street 12
Saving the world 12
Returning at night 13
Election 14
Tiger 15
Don't smile please 15
Sweet tooth 16
Fears 16
Low-key haiku 17
Wheels 18
Supervision 19
Ageing writer in a pub 20
Miniature modes 21
Memory 22
Servicing 23
An old man reminisces 23
Consumer society 24
As seen on TV 25
The old stories 25
Envy 26
Rights 26
Tribute 27
Cincinnatus 27
Yusoff the bold 28
Thinking of the young ladies of Tsuda College 29
Old-world 30
Castles 31
Going wrong 32
GP 33
Shop 34
Seasons 35
Forgetfulness in Fuzhou 35
Self-criticism 36
Cat dream 37
Emperor dream 39
An even more ludicrous dream 40
First words, last chances 41

A quiet neighbourhood 45
Pensioners at the post office 45
Citizen's Charter 47
A half-remembered tale 48
The bright side 49
Actually 50
Traffic 50
A quandary 51
All the things 51
Confetti 52
Dream snatches 53
Warnings, warnings! 57
The great man can be quite funny, nicht wahr? 58
'You're mine thereafter' 60
As if human 61
Bindings 62
A register of diverse deaths 62
A book at bedtime 63
So many 63
What we never and always think of 64

'Old men and comets have been reverenced for the same reason: their long beards, and pretences to foretell events'
—Swift

Clichés

Thinking of my grandmother
And a loving gesture of hers—

Then seeing it is my mother
I am thinking of.

The two alike in more ways than one,
Both of them dead.

Thinking of things like this,
Then sighing: 'But it's just a cliché.'

For the old most things are platitudes;
Knowing it makes no difference.

Coming true

Clichés (how one used to despise them!) are coming true. The sun rises, and the sun goes down. There is a season for everything; also for nothing. The knees tremble; desire falters (except for a quiet drink). The grinders aren't what they were (they were never much). Can't see too well; soon there will be an end to the reading of books. Nor hear: the daughters of music are brought low on Radio 3. You look on your labour, and mostly it's vanity or vexation. There is no remembrance of things past. Fears are in the street. Jugs and cups get broken. No doubt crickets would be a pest, were there any left. Yes, the words of the wise (God damn them) are like nails. One nail drives out another. So find some better clichés.

In the street

Did I imagine that romantic story?—
England 1919, and the war just over,
It was raining hard, and she could see
A soldier, looking lost, was getting wet.
Her umbrella offered decent room for two:
And that was how they met.

He didn't rejoin the Dublin Fusiliers,
Didn't go back to Ireland,
Little work there, lots more rain.
Better to stay and be a British husband.

Did our mother really tell us this,
Or does remembrance misconstrue?
She was never given to romancing;
Either way it could be true.

Saving the world

At Christmas our father took us to his church,
The Catholic, though he only went there then,
When he thought we ought to see the famous crib,
Its painted figures of animals and people.
I felt at home in that foreign place, the scene
Reminded me of Noah's Ark, my fondest toy,
Where the animals went in two by two, or
Sometimes one by one, I didn't always count.
A story lay behind it, how the world was saved,
Animals and people, by a new beginning.
(Which was why there needed to be two of each,
Like my two white mice who created a nation.)
The Nativity at St Peter's was much the same,
Except for having just one baby, fast asleep.
But by next Christmas there would surely be another,
Then the Ark could come and carry them away.

Returning at night

Walking homewards, after a drink with the boys
(And you the oldest of them nowadays),
Down the dark descent from the station:
Raised to your full height or rather more,
Fearsome of mien, your pipe clutched like a pistol,
Mumbling obscure threats, you not only discourage
Muggers, you look very much like one.

Then striding onwards, across the final busy road,
Quite unconsciously, until you brush the kerb . . .

Perhaps a car hit you? (Daredevil drivers
Round here, and you the poor devil.) Yet you go on
Walking, muttering, scowling, into the house,
Hailing your wife: dead but you won't lie down?
And after this brief zombie interlude,
You'll find yourself in dreadful pain, in hell,
Or else in dreadful peace, in heaven?
(Unless, of course, in no place, feeling nothing.)

'I was expecting you,' the missus says. 'Supper's ready.'

Election

No longer able to tell one from another,
Or the real face from the caricature . . .

But he trundles off to the polling station,
A church round the corner.
They have screened off the altar,
Jesus isn't standing on this occasion.

He stares at the ballot paper.
'What! No NSPCC, no RSPCA,
Not even RIP to vote for!
It's a crying shame, a stigma,
That I should live to see the day!'

They ask him please to leave. Such language!
(Drunk, at this hour, at his age.)
But first he'll raise his eyes to heaven:
'The only party I would ever trust
Is one that doesn't want to govern!'

They exchange small smiles
And tap their heads discreetly.
As he perceives, and chooses to ignore;
He has done his duty.

Tiger

What a good-tempered old fellow he is!—
So it strikes him, watching the soap operas,
Coronation Street and then *EastEnders*.
People nagging and wrangling like crazy—
What a sweet nature his must be!

Yet that's life, isn't it, all that hubbub?
He had better find someone to squabble with
(Not someone who might forget the supper
Though, or leave the bed unmade).
'To the Editor: I shall call a spade . . .'

Don't smile please

Since the primary school is next door
You can't help passing the playground
But don't you smile at the children
Whether a small girl or a little boy
Don't you even look
You know what people will think
And you really can't blame them.

What a world we live in! What went wrong?
If there's another world to come
Let's hope it's one where people smile
And you can smile back safely.

Once they asked you to return their ball
It had sailed over the palings—
Eyes cast discreetly upwards, you stepped
Into the street and were nearly run down
Still, a little boy said 'Thank you, mister'
A small girl almost smiled.

Sweet tooth

You take to what you never cared for:
Sweet wines—
Come back, Paradiso!
Or what you only liked when young:
Sweet cakes and puddings.
Well, your teeth aren't likely to suffer,
And if you put on weight, what matter?

Something to sweeten the sourness of age,
Its waning appetites.
Be bold: tuck into Terry's All Gold,
Swill a glass of oloroso!
(But wash it down with fino.)

You see through a dark glass.
You speak as a child again,
Putting childish things before you
(Behind you the Apostle Paul's esteem):
'A nice large helping of ice-cream!'

Fears

It's nice and peaceful. Peace is rather nice.
What's that? Loud noises in the middle distance.
Yobs on the staircase, housebreakers in the hall . . .
Strange sounds, hardly human, not hate, not frenzy.
Martians bursting through the ceiling?

No, it's faint noises near at hand . . .
It's—yes—the stomach, fancy that.
Mildly remonstrating, or perhaps applauding.

Low-key haiku

Films that don't reveal
Their titles till halfway through—
    Much like life itself.

So long to learn! And
Short the time. And when you've learnt,
    'Times have changed!' you hear.

Laying bare the soul
Which he isn't sure exists—
    Such sad nudity!

The glory that was . . .
Pillars, battlements, turrets—
    Now fallen arches.

The stand-up comic
Needs to sit down—he's as old
    As his jokes, he says.

So easy to grasp
Though once unfathomable—
    The medical books.

Obituaries—
You scan them as though you were
    Part of a tontine.

A kind of Velcro
Barely holds them together—
    Your body and soul.

Wheels

On your 70th birthday, a card from a
Far-off unforgotten land, which reads:
'Forgive your friends!'
Signed by a former Minister of Culture
(Retired from office, he cultivates his garden),

Who once had threatened deportation
In terms which would have been quite wounding
Had you been thin-skinned,
And even so were disconcerting:

Beatnik, mendicant, passing alien—

Who had you banned from press and radio
Though not—your own small garden—
The insignificant classroom.

Friends and forgiveness:
Such grand words, thirty-odd years after!
A creaking wheel has come full circle.
One still stands where one stood,
Or sits, faintly bemused, a trifle giddy.

Supervision

Below an essay on Shelley he wrote:
    'I don't think we're here to judge his soul.'
A telling reproach, whatever one's view of souls.

A fine teacher! He knew the proper medicine.
Self-righteousness would never be the same,
It ceased to be a right.

He could never keep his pipe alight,
Smouldering matches rained about him.
Once he gave it up, to discipline the spirit.
His auntie told us over tea and cake:
    'Because he burnt a hole in his trousers.'

Sound on poetry: a major work on Wordsworth
Was diffidently in the offing. When one of us
Favoured Shaw's over Shakespeare's *Cleopatra*
He dropped his pipe in wordless pain.
But the soul was what most concerned him
(Though he chose not to dwell on it),
Which we are not here to judge.

Something we learnt from his auntie too,
Who baked the cakes and supervised his body.

Ageing writer in a pub

'Art's long? Can't say I've noticed it.
Books have a short life, let me tell you.
One day you're admiring your advance copies,
The next, the whole lot's remaindered,
Gone like a pint of beer—ah, thank you!
Or *put out of print*, which sounds as if
The vice squad are feeding a bonfire,
Or a vet's dispatching unwanted pets.
You'll be hard pressed to find a backlist
Now, it's all vanguard, I guess, and no rear.
Still, thinking of the books that perish young
Does make you feel how long your own life is . . .
And there's a book one might have a crack at,
The Dotage of the Adage. Great subject, eh?'

He starts to scribble on a paper napkin.

Miniature modes

Hard-headed tanka

   It's a work of art
It's a bestseller also
   Works of art alas
Remain imponderable—
Selling best is palpable.

Soft-hearted tanka

   A daughter complains
An ex-wife is complaining
   Bitching makes big books
Page after page of paper—
You can't hear trees complaining.

Sighku

   Like tapioca
(From Tupi for 'squeezed out dregs')—
   These modern novels.

Unheroic couplet

In antique times men scarcely knew a dustbin,
They had so very little dust to put in.

Ancient saw

If only youth knew, then it wouldn't do.
If only age could, most likely it would.
Some deficiencies partake of virtue.
Ignorance and impotence both bring good.

Pieku

   The perfect pork pie?—
Guzzled by the ideal
   Woman in the sky.

Memory

When she read what I'd written,
Verses about those early years,
I saw my mother's lips tighten.
'It's dedicated to you,' I started to plead,
But that was no excuse.
'You've got a good memory,' at last she said.
Something, but not exactly praise.

To have good memories was pleasant,
Not so to have a good memory.
There was something faintly indecent
About it, it got you a bad name.

Reading Proust, I see how she felt.
A good memory: one of those luxuries
Not meant for the modest likes of us.
It was rude about others, it dwelt
On what was over and done with,
And it sapped your energies.

At last I grasp her ancient statutes.
When little is left of the flower
You revisit your roots.

Servicing

How distinguished! To be examined
Fully dressed at a discreet distance,
Stock-still, like the lady of Amherst,
Sworn foe to all familiarities!—
Instead of stripping down to your smalls
(Modelling for St Michael, or Sebastian?),
Tendering your ambiguous goose-flesh,
Then struggling back into your clothes—
Simply to hear the foregone conclusions:
You should give up smoking, cut down
On drink, take exercise, lose weight . . .
You seem to have lost a shoe already,
And you have taken exercise.

An old man reminisces

'A gruesome sight it was, I can tell you—
Black with them, the length and breadth of it,
Scores of small bodies scattered all about,
At every angle, some still twitching . . .
There was no choice, it was necessary,
But I had to turn my eyes away . . .
You don't see many of them these days,
Once in a while a lone, bemused survivor,
Promptly polished off by those new devices,
Flitguns they call them. Few fly-papers now.'

Consumer society

(with acknowledgements to Daniela Crăsnaru)

When we were children we went to the pictures
And saw a starving man who had a vision:
His friend turned into a juicy chicken.
How we laughed! We knew he wouldn't eat him.

*

One poet jokes to another: 'Poems get you nowhere.
If you want to be famous in Europe,
Truly famous, you have to be a cannibal.'

A few years pass, and a mother looks at her child,
Remembering what her fellow-poet had said.
She could be famous, she thinks. (And less hungry.)
'Fame is knocking at the door' in Romania.

*

And now, with supper sure to follow,
We attend the cinema, to observe the species
Of murderers who consume their victims.
They are famous in Europe, in America too.
To what large feast is this an appetizer,
We might ask. Who will eat, who be eaten?

As seen on TV

Someone will be rapped over the knuckles
For a rape too close to the watershed—
The kiddies would hardly be out of the room,
Only halfway up the wooden hill to Bedfordshire.
After nine o'clock however rape is admissible,
Grown-ups can relish it in the privacy of their parlours
Exempt from the questioning of untimely children:
'Doesn't the lady like what he's doing? You don't mind,
Do you, Mummy?' and 'Why has Daddy gone all red?'
They must have seen something nasty in the watershed—
Someone should be given a good talking-to.

The old stories

How they bring a tear to the eye (or somewhere near), the old stories! The things we used to believe. Thus, that evil spirits could do no harm in the presence of a pregnant woman, since she bore within her a being wholly innocent. (Hence such women were much in demand at conjurations and exorcisms.) Nowadays there is no guarantee of innocence, nor is innocence a guarantee of anything. But just imagine, in a crowded tube train: 'Madam, pray take my seat, I insist, for inside you lies a creature that, unlike the rest of us, is one hundred per cent guilt-free.' And imagine—oh, imagine all sorts of things . . .

Envy

How their ways prosper!
But they will get their comeuppance
One fine day.
The triumphing of the wicked is short.

Yet success succeeds success,
Their bulls gender, their cows calve, their books sell,
Their triumphing is long.
Their sins do not find them out.

How the heart sinks!
Until you think: but I too
Have been spared. Not all my sins
Have found me out.

Though there's no justice in the world,
There's an injustice that can work as well.

Rights

These days he was readier to take a tip.

'The right of God to be identified as Author of this World has been asserted by Him in accordance with the Copyright, Designs and Patents Act 1988.'

It was an act He hadn't thought of before. He liked the word 'asserted'. He hoped it wasn't too late.

Tribute

Many writers prefer music however—

It bypasses words, it means what it is
Is immune to original sin
Uncorrupted by evil communications
Never liable to solecism or delusion
It carries love, it can do without sex
It doesn't stop to think
As it were defrosted architecture
It is all of heaven we have below
(And all art constantly aspires to its condition
Aside perhaps from the poems that aspire to
   the condition of prose)—

Also it provides an agreeable background
To their lowlier labours.

Cincinnatus

They found him in the fields, told him his country needed him. In the capital they had drawn up a short list of likely saviours. Not so short. Crowds of farmers large and small gathered together, worrying over the weather. Interviews and deliberations went on so long the crops failed. The country starved, the people rioted, the army mutinied. In despair they advertised for poets, novelists and playwrights.

Yusoff the bold

Once again one thinks of
The former student Yusoff
And his courage in the face of—

The very face of Mr Lee
Prime minister and stepfather of his children
Who came to inspect the university
Stalking through its academic Eden
Like a dyspeptic tiger
In hot pursuit of utility
And the filling of the national belly

What role will they play in the future?
Student after student
Wisely answers 'Teacher'
Until it comes to Yusoff
(The race is to the swifter race)
Who bleats 'Er, me?'
Like a goat face to face
With a tetchy tiger—
'A poet'

An awful silence
Except for the stifled gasps of the retinue
It is seldom that Mr Lee Kuan Yew
Encounters such impertinence

Yet let it be told—
Those who avowed an intention to teach
Have become administrators
Administering whoever is left to teach
While Yusoff the bold
The quondam tethered goat
Goes his way as a poet.

Thinking of the young ladies of Tsuda College

(after a Japanese poem)

A customer in the bookshop has a grievance.
When she asks for the poetry corner
An assistant smiles and shows her to it,
A shelf containing books about embroidery.
Shishu means 'collections of poems'
And, by homophony, also 'embroidery'.
Understandable, must happen all the time.
(Better than getting *Pavane for a Dead Infanta*
When you asked for *Songs on the Death of Children*.)
But why, she wonders, does the male assistant
Assume that what she must want is embroidery?
'Because I am a woman,' she starts off,
'You suppose my only concern is with—'
Then stops. There are other battles to fight.
They could put the embroiderers all in prison,
She muses, yet embroidery would still go on,
If only for the sake of women's underwear.
Shishu and ditto, in different characters, *shishu*.
She asked for a corner, and now she's in it.
There are tighter ones, and narrower escapes.
Best simply to thank him, and pick out something
To buy, like *Embroidery, Our National Heritage*.

Old-world

Having arrived at wherever it was
The car stopped, and you hurried round
To the other side and opened the door.
Whereupon the driver fell out.

When she had dusted herself down
She said it had never happened before.
Perhaps she hoped it never would again,
But she seemed obscurely affected.

Castles

There is always a castle, or what passes as such,
Where strange things happen, around which
Sinister stories abound. A place to be avoided.

And of course there are peasants, there have to be,
And one is generally one of the peasants—
Trusting that folk wisdom is occasionally wise:
Turn your face to the wall as the horses pass by.
Keep clear of them, child, for nothing but shame
And sorrow can follow! Don't speak to strangers,
Never believe the gentry, never trust officials!

Dim shapes are glimpsed at times, engaged in obscure
But horrid acts. Cries are heard, hard to interpret:
Their pains are not ours, nor are their pleasures.
One shakes one's head, one mutters darkly,
The innkeeper seals his doors, the priest his lips.

What was it scared old Joseph out of his senses?
(Or was he born a halfwit? Nobody remembers.)
What happened to Ebenezer, goatherd and bellringer,
Who walked abroad one night and vanished for ever?
(Or did he leave to join a distant daughter?)

Once a lady came, to write her poetry, she said,
In peace, in the truth and beauty of nature.
A nervous lady, not cut out for these parts,
Thinner she grew, and paler, scared of everything,
Of her cat itself as it purred by the fire.
Suddenly she was gone, the cat circles the village.

One does one's best not to huddle in doorways,
Not to study the moon, or hug a crucifix,
Not to leave too many sentences unfinished.
Sometimes one whistles a guarded defiance.
Will our children be bolder and rise up in anger?

But there will always be some sort of castle,
Remote, but too close for the comfort of peasants.

Going wrong

If only they hadn't made speeches!
We mightn't have been so bad
Except they made speeches. If they hadn't
Reckoned we were so utterly different
We might have been a little more similar.
But no they made speeches
Sometimes with sticks held behind them
Or sweets tucked into their pockets.
We sensed the sticks, we smelt the sweets.
Maybe there was good in us, they knocked it
Clean out of us, we went and did otherwise.
The speeches showed us how to.

You said that we ought, you knew we wouldn't.
If you hadn't kept talking perhaps we'd have
Listened. If you hadn't made speeches
We mightn't have made trouble. We could have become
As good as you were. We might have been better.
Was that why you went on making those speeches?

GP

General paralysis of the features, you muse
As you enter the surgery. 'Yes?' he asks,
Long-suffering. 'What seems to be the trouble?'
You know not seems. But you don't say so.
This isn't the time or place for badinage,
So don't go quoting Proust on medicos.

'I could give you pain-killers,' he allows,
And lists his reasons for declining to,
'Side-effects . . .' Then, 'Let's see now,
How old are you?' As if he didn't know.
You are a number, a number of years,
And there lie diagnosis and prognosis too.
Age is a side-effect of youth.

Poor fellow, he's so bored. Cheer up,
You want to tell him, there are hordes
Of hardy kiddies in the waiting-room,
One has just thrown up on your magazines.
You only say: 'Goodbye, doctor, thank you.'
He smiles at last. At least you haven't argued.

Shop

'Curious,' remarked the pain, 'a man's body, there's so much of it.'

'A woman's has more to offer, I always think,' said the other pain, feigning a sigh. 'Trouble once a month at least. And sometimes in her ears. I'm saving her breasts for later . . .'

'The big toe,' said the first, 'the knees, the back . . . the teeth alone are enough to keep me busy. You wouldn't believe it.'

'I would, I would. Only the other day she caught a fingernail on something, the fuss she made! Calling out for you know what.'

(For what were called pain-killers.)

'And the stomach, a study in itself,' said the first, 'and the head, and the chest—you name it . . . All this travelling does take it out of you.'

'Good for the health, though,' the other said. 'And it's important to have a purpose in life.'

'To tell the truth, I'm bored with the fellow. Been up and down him, the whole length and breadth, time after time. Where's the challenge?'

'Wouldn't blame you if you gave up on him,' said his colleague. 'Sounds like a dead loss to me.'

'True, there's always some body else.'

Seasons

One sentence in English he knew by heart:
'If Winter comes, can Spring be far behind?'
It sounded cheerful; it usually fitted.
He was a writer. He had translated *Quo Vadis?*
From the English. What else he had done
We never learnt, nor what had been done to him.
Plainly he'd had a number of hard winters
Known choicely as the Cultural Revolution,
Made to clean out latrines, at very least.
If you think that's a doddle, just the job
For spoilt intellectuals, then go and look at one.
Winter went, spring returned. Just the two seasons.
Once I think he started to say 'If Spring comes . . .'
But his English failed him. Or someone was listening.

Forgetfulness in Fuzhou

'Ah,' he declaims, pointing across the Strait, 'Formosa!' 'What did you say?' snaps the interpreter, a severe young woman. 'You mean Taiwan. You must not say Formosa, it is an imperialistic name, it is not ours.'

No doubt. But a lot has happened since, he thinks, other imperialisms, and worse. If he mentions Tibet he'll be made to climb some huge sacred mountain. 'The Portuguese word Formosa means shapely, beautiful,' he murmurs piously.

'Oh?' The chill lifts. 'Taiwan is indeed very beautiful.' (And no doubt will soon be hers.)

She favours him with a smile. At times these old foreigners talk good sense. Or—which deserves respect—come up with a good excuse.

Self-criticism

Esteemed fellow-workers and in some cases friends
It is gracious of you to gather to hear and help me
In my hour of need of helpers and hearers.
I come to confess my failings and felonies
Errors, backslidings and premature forwardness
Sins committed and some of omission—in groupings
Regardless of gravity and open to addenda.
When the tongue begins to trip, the words must fall.
From no great height to start with I have fallen far!
(And what, you may ask, is the function of criticism?
A very good question, we shall leave it till later.)
But now let me itemize as best or worst I may
In this unsweetened session of full-throated thought
Sinkings into iniquity, dire deviations and derelictions
Trespasses and transgressions, slips and solecisms
Exactly as alas they happened, step by step.
Plain speaking among comrades is only proper . . .

. . . for evil communications call for fearless ears
And truth we are told lies in a bottomless well.
Make no mistake, my every misdeed one by one
Shall be dragged protesting into the light of day.
True it is that hardened cases make good law!
The function of criticism remains to be probed still
But the day is fading, my friends, and the canteen
Is about to close. I have delayed you too long.

Cat dream

'I'm happy to see you all again,' I say,
'Even though you happen to be dead.'

Prompt the response, perfect and plain-spoken,
If a trifle shrill: 'Pray watch your language!'

Ten lives, then? Timidly I ask,
'Where are you now, when you're not here?'

'In heaven, of course,' their elder says,
A cat I've known for many a year,

'Which ninety-nine per cent of cats prefer.
The residue, I fear, are hell-hounds.'

'It's really good of you to call on me.'
At least I hope it will be.

'We had to, you could hardly visit us,'
Comes from a tabby in sardonic tones.

'Was it in heaven that you learnt to talk?
I mean, to speak such fluent English?'

'We always could, we speak in tongues.
Simply we had no wish to discompose you—'

'Hier spricht man deutsch,' a polished Siamese
Puts in, 'Ici on parle français . . .'

'—Better to rest content as your dumb friends,
So you could find no ways to make us talk.'

'Somehow you look less innocent,' I venture.
'Where we live now, there's little need for looks.'

(Albeit she's kept hers, insinuates
The Siamese through sly angelic eyes.)

'Your manner, so it seems, is more severe.'
'Nor is there any need for softness either.'

'No pets in heaven!' the tabby hisses.
The elder chides: 'All are joint favourites there.'

('Soft purrings,' sighs a prudent black,
'Turn wrath away, where'er you be.')

'A rotten job you made of it,' the tabby
Grouses, 'when you dug my grave.'

'My heart,' I plead, 'just wasn't in it.'
'My body barely was,' he growls.

I beg forgiveness for outliving them.
'But have you?' is the question. 'Are you sure?'

'Hush now!' the elder says. 'Fair's fair.
It is his dream, remember, we are guests.'

Then they unbend a little, almost smile.
They purr their brief farewells, and fade away.

Emperor dream

It didn't seem at all odd that they should make me Emperor—of Japan as it happened. Or that at my side stood a lovely young consort, her features partially obscured under a light veil. (I was saving her up for later.) Dressed in elaborate ceremonial robes, we walked graciously among our people. I would be good to them, they would be good to me.

Three large ill-favoured men came up, saying I was in grave trouble, I had been taking notes of the secret sacred ceremonies. Certainly not, I replied with dignity, I had done no such thing (though admittedly I was thinking of writing a piece on the subject afterwards), and if I had—well then, it was my ceremony, it was my coronation. They scowled.

I would walk among my people, they would cherish me as I them. I sought for comfort the delicate arm of my gentle helpmate, and clasped a hairy tree-trunk. It was the arm of a muscular transvestite, a hefty hermaphrodite from the gay quarters, a police spy, an assassin, a hangman. My robe, I noticed, looked much like a crumpled dressing-gown, I could see my pallid shins, my bedroom slippers. These were not my people. Probably I had no people.

When I woke up I was still alive. I set down these events at once.

An even more ludicrous dream

Outside, a battle. Stylized confusion
As if on stage, dull thuds, sharp cries,
People killing people in profusion.
Suddenly it stops. A truce is called.

Inside, a conference table. And me,
A responsible official. (Rather unlikely.)
But what is this? My buddies embracing
The enemy, who appears to be a Swede!
(Most unlikely.) I am righteously appalled.
First comes killing and then comes backslapping—
What was the point of it, what the need?
I fling down the piled-up papers I bear
With an almighty thump at his feet
And stalk out, nose in the air.

At once I begin to worry.
I may well have committed treason
Or should it be mutiny?
My best friend won't tell me.
I think, the Swedish officer can't be faulted,
It was somebody else entirely.
But he was the one I insulted.

I shall beg forgiveness of the Swede,
My one and only apology—
So I assert, but they pay no heed.
It isn't the slapping of backs
I'm against. Or other tokens of amity.
I only desire to make my position clear,
It's the absurd anomaly.

Peace has been shattered, I fear.
Thuds and cries. And I am an outcast.
Opprobrium spreads through the night.
I wake up sweating and aghast.

First words, last chances

   Words you've never used
And have always wanted to—
   Get them in quickly.

      *

   Dight in dimity
Enlaced with lazy-daisy
   In fishnet fleshings.

   It fell on your head
Her old boyfriend's framed photo—
   Fearsome xoanon!

   The ergonomics
(Please don't tread on erica!)
   Of the percheron.

   How the aasvogel
Flapping alongside the jeep
   Churned up your bowels!

   An ancient bayou
A batrachian flops in—
   Sound of H_2O.

   By the billabong
A sheila throwing goolies
   At perving galahs.

   Jalousies muffle
Criminal conversation—
   Discalced and unfrocked
Ithyphallic, perforate—
A case of jactitation.

Being discovered
In flagrante delicto
With a young person
In statu pupillari
He pleads hypnopaedia.

Dreaming you're starving
Lost in a forest: millefeuilles
Drift down to your mouth.

Sucking at jujubes
(Junketing, jabberwocky!)
Sipping mint juleps.

It still disturbs him
The tintinnabulation—
Sunday, odd day out.

The toilets are closed
But there's a dark passage near
Known as Pis Aller.

Membrum virile?
It matches the mise en scène—
Ah, micturition!

A night of mixed drinks
Nepenthes and mnemonics—
A crapulous dawn.

What Tagalog means
Is neither Tag nor Logos
But 'native river'.

Godthaab thought it best
After twelve score years and ten
To call itself Nuuk.

She dropped a taipan
(Oh what a kakemono!)
On her tatami.

The term congeries
Given by Gavin Ewart—
    Jellied eels I'd thought.

Zoomancy (four
Syllables please note) as when
    Cats augur earthquakes.

Kin to yin and yang
Lingam and yoni are less
    Metaphysical.

Odalisques, frillies
Frou-frous—and aboulia:
    Let's mithridatize!

Everest, Mont Blanc
Matterhorn, Mons Veneris—
    Hills so hard to climb.

Eleven ages—
Nonage, bondage and rampage
    Marriage and mortgage
Hostage, wastage and outrage
Dotage, wreckage—average.

An opsimath digs
(Paradise or parados?)
    His hortus siccus.

The best transhumance
(Tsunamis are predicted)
    Calls for theurgy.

Your present prospects
No longer quaquaversal—
    Mere tunnel vision!

A biography—
Asterisks, obeli, but
    Where's the bloody text?

All in all he was
(Here's a pleasing epitaph!)
Alphanumeric.

Vox angelica
(Voicing vale or ave?)
Or vox humana?

Silence draws closer—
What chrestomathy helps you
Learn a tongueless tongue?

*

But they didn't stop!
Nearing heaven, nearing hell
Babel's still building.

A quiet neighbourhood

Shocking. On the street, a man intimately and ardently embracing a woman, the woman forced back against a garden gate, her head lolling over it. They don't seem especially young. Passers-by try not to look; and look. The man is groaning. His wife has had a heart attack, he is desperately holding her up. Both of them are distinctly old now. Someone who was about to call the police has a change of heart and rings for an ambulance. Such a quiet street.

Pensioners at the post office

'Nice weather for the time of year, considering. I see you haven't got your stick today.'

'Considering what? Must have left it in a shop.'

'Oh dear, I've written in the wrong place. Do you think I'll get into trouble?'

'Where are the grandkiddies? Not outside in the street?'

'Did you see that enormous dog? A ballpoint terrier I wouldn't be surprised.'

'Our cat doted on my hubby before he had his leg amputated. My hubby, I mean. After that the cat couldn't stand him.'

'Can he stand himself? Oh er, I didn't mean . . . I think I've put the date wrong.'

'I never feel safe when I've just collected the pension. These muggers, you know, they can tell.'

'That lot, what are they called—Securicorpse—they ought to see us home safe . . . Where's my bag gone?'

'You're holding it.'

'I've been and spoilt the form. I hope they won't be angry. You know how fussy they are.'

'He isn't *they*, he's an Indian.'

'When he's behind that counter he's *they*. I've always been nervous of them.'

'Indians?'

'No, *them* . . . They're a bit off-colour today, poor little loves. Reckon it was one of those take-aways, you never know.'

'Cats don't care for change. Like amputations. Or new shoes.'

'But if . . . Oh, I wonder if the camera's on. Just fancy, us being filmed!'

'I think it's a liberty. In my old coat too.'

'Of course it's all videos these days. We don't have the machine, not that I miss it at my age.'

'Dogs don't mind. They're not so particular.'

'Perhaps they'd be pleased if I bought a stamp or two as well.'

'That woman's buying an awful lot of postal orders. Wish there was somewhere to sit down.'

'Won't be long now . . . Will you be going my way, dear?'

'Only if I find my stick.'

Citizen's Charter

Cemeteries are grouped under Leisure Services
For you will have lots of free time on your hands.

You are given a choice of future addresses
(E.g. Putney Vale, Magdalen Road, Battersea Rise)
And assured that your religious beliefs, if any,
Will be duly accommodated as far as is possible
(With God all things are possible, not so with Local
    Authorities)
You may desire to be strewn in the Garden of Remembrance
    by appointment
And to have your name artistically inscribed in the Book of
    Remembrance
A Burial/Cremation Officer will attend, to ensure
That the ceremony meets with your entire approval
A peaceful and pleasant environment is guaranteed
Grassed areas are cut regularly and flowerbeds weeded
And dogs are forbidden to exercise without their owners
Should you have entered into a Maintenance Agreement
Applicable problems will be fixed within two working days
    at the latest
Please make use of the bins and report any vandalism
    at once
We are committed to dealing with all complaints in an
    understanding way
But if you are still unhappy you should approach the Chief
    Executive who will take an independent view
It is left to you to find your own means of communicating
Whether by appearing to the Deputy Principal Cemeteries
    Officer
At dead of night, or haunting the Town Hall, or interfacing
    with the Council's computers.

You may have made plans for your retirement, but are
    reminded
That the afterlife is not necessarily a bed of roses.

A half-remembered tale

'A day's outing,' they said smiling. She knew about outings. When she was a child her grandmother went out for the day, and never came back. 'Up a mountain, I suppose?' she was about to whimper, when one of them said, 'Up a nice mountain, dearie.' Wouldn't she like that? She wouldn't, it was getting cold out there. 'The last chance for a little trip,' they said, 'before the freeze sets in.' People were busy round the picnic hamper.

It was a long outing. It was a steep mountain. But they were very helpful. 'Take my arm.' She almost said, 'I'd rather have your legs.' At this rate they would hardly get back home by nightfall. She certainly wouldn't.

Then someone stopped, the young one who had been to school, and thumped the ground with his staff, and said something like 'But this is a stereotype!' Whatever that might be, it sounded bad. 'Must be some nasty foreign habit,' one of them whispered uncertainly. 'Oh dear,' they muttered one after another, shaking their heads, 'it won't do, will it?'

They carried her down the mountain. The wind was fierce, there was snow in the air. They moaned and groaned, but she could tell they were glad to avoid that scandalous thing, a stereotype. 'After all,' somebody murmured, 'it can't be for long, we can just about manage . . .'

The bright side

Awaiting news, waiting anxiously,
Wondering what and how to propitiate . . .
But now, though the waiting's no tougher,
If it's bad, there's time enough to suffer,
If good, scant time to profit plenteously.

The thought provokes a smile, hastily
Disowned but hard to extirpate.
Old buffers shouldn't need a buffer—
And yet it helps you while you wait,
Anxiously, and now a mite shamefacedly.

Actually

'Actually,' says the four-year-old grandson . . .
Pardon? 'Actually,' and goes on to explain
That what he has just drawn is not merely a car
But a BMW, 'actually'.

'Actually'—you think of telling him—wasn't
Heard of in your youth, nor were BM-whatsits
(There were Hercules bikes or Raleighs
Powered by healthy legs, or bloody aching ones).
And actually—you almost say—his life
Will be a darned sight easier than yours was;
Except it might be harder, actually.

But in fact you say nothing. Or only:
'Actually that's a fine motor car!'
And you win an approving glance.

Traffic

Hearing that he's to have a sibling,
He will feed it, he declares, and bath it,
Dress it, and take it out for walks.
(Walks of course are the most important.)
A methodical child, he resolves to practise,
Instals his teddy as a fitting proxy,
And drives the pushchair down the street.
No one has warned him about pedestrians,
He only cares for Baby Bear's well-being . . .

The hungry generations don't just tread you down,
They run right over you.

A quandary

'Marmite is *good* for you,'
The boy announces smugly,
Meaning him. He is very hungry.

'So it's good for me too?' you ask,
A wolfishly grinning grandpa.

That's a tricky proposition.
You are not he, you belong
To another species.
Marmite could be bad for you,
It could make you very sick.

And if you hear it is good for you,
You might lay hands on his sandwiches.
And he happens to be very hungry.

It's one of those foolish questions
Best left unanswered,
Even though silence goes against the grain.
He stuffs his mouth with bread and marmite.

All the things

He will be left on his own for a while
For there are things one needs to do.

'I shall sit by myself,' the boy concurs,
'And think of all the things I know.'

One has forgotten many things one knew,
Others one prefers not to think of.

Luckily there are things one needs to do.

Confetti

The grandson's mother, alias his daughter—
He recalls the night she began to be born.
Feeding a pile of saved-up old pennies
Into a slot in a freezing phone box,
Press Button A, or maybe Button B.
At last there's a potential taxi in Brum.
At last an actual one arrives. So he shoves
His wife inside. It's stuffy, sweetly scented,
And beery. And it's littered with confetti.
Attendant ironies? The cabbie doesn't notice.
He's weary. It's been a long day. He's wrapped up
In something. Or rapt. Perhaps the wedding was his.

Dream snatches

Your mother, young, smiling, in a cloche hat. An old photo you came across yesterday.

The milkman goes past. Naturally he carries a cow on his float.

You are writing like Henry James, but better. Splendid stuff, and lots of it. (Where did it all go?)

A fearsome she-vampire asks sweetly for a gin and tonic.

A blazing day on a tropical campus, searching for your students, students of twenty, thirty years ago.

You are paying your last respects. In Chinese fashion, the Vice-Chancellor's coffin is open on the floor. He starts to talk to you about some boring academic business. You bend down and tell him: 'You should speak to your wife. She's dreadfully upset.'

An acerbic old boss, ten years dead, is being enormously kind.

Poems, typescripts, notes left at home, train missed. You find a phone box. It's full of incomprehensible dials and buttons.

In a dull dream one is relating one's exciting dream.

The young princess is about to fall off a large elephant. You will catch her. No doubt she will want to marry you.

There's a vile monster at the bottom of the stairs. In terror you fling yourself down on it. To be continued.

One is giving birth. It's extremely painful. One feels sorry for women.

Nonchalantly piloting a plane. All's going perfectly. Then you realize you don't have the faintest idea of what to do next.

You know your old friend the sensei is dead. But you are with him again, turning Japanese verse into English. You ask if he's seen Issa. Ages ago, at the eel restaurant. But Hokusai dropped in the other day. 'Look, he gave me a sketch.' He points to the wall. You see nothing there. The wall fades. Everything fades.

The woman's there. Then not. Slipped into another room, crowded with people. There she is. Then slipped away into a crowded room. Then gone again. Into another crowded room.

A sick old cat is chirping: 'You worry too much.'

A huge cylindrical parcel arrives, rolled up like a carpet. It must be in response to the typescript you delivered recently. It has been in the rain, you can't unwrap it, the sheets are stuck together, the ink has run.

Looking for a lavatory. Here's one at last. You rush in. You are on a lighted stage.

The three of you are to be taken out and shot. 'I'll go first,' you say, getting to your feet. Because you couldn't bear the waiting.

Your daughter has a baby girl. In the morning your daughter has a girl baby.

One is caressing one's new cat. She has two lovely white breasts. The next day one learns she's a he.

A girl is desperately ill. You save her life with a timely injection. But how to hide the syringe? You will be accused of peddling drugs.

You make a truly brilliant and original pun. You laugh so hard it goes out of your head.

One is an iron penis. But one can't find an iron vagina. Just as well.

One's publisher has gone bankrupt. One feels terribly sad.

Talking to policemen. You admit you have killed someone. For utterly sound reasons. Which you can't recall.

Sunlight, weeping willows, green green grass, people dancing gently in a ring. Pure loving-kindness. The like of it you have never known; but you know it.

Lee Kuan Yew pleads for your help: he is on the run from his enemies. You will help him.

In a stuffy apartment someone is chasing you round a massive table spread with Turkish delight. It must be 10 Rue Lepsius. 'Cavafy!' you gasp, 'You ought to be writing your poems!'

J. B. Priestley is saying: 'I can teach you how to write a comic novel.'

A fervent young woman proposes that permission fees be paid to persons who feature willy-nilly in others' dreams. You look around nervously. Dreaming above your means?

You don't know how you got to the top of this tower, nor how you will get down. Except by falling. Which you do. It's like dying.

One has dreamt it so often one knows it step by step. When it begins one cries, 'It's only a dream!' The dream pays no attention.

All night long you agonize over two words and their correct order.

You are not surprised by your breastplate and long sword. 'This is the Middle Ages,' you remark cheerfully.

The flight has arrived. Or departed. Either the other never arrived or has already departed. The lights go out.

The dead man rises up. Immense relief. He bursts into flames, suffers horribly. He dies. Immense relief.

High on a hill, precise and detailed, York Minster. (Which you have never seen.) You're standing on the very top. The Devil is offering you a sizeable tract of freehold land. You feel confused.

Having beaten you easily in a cross-country race, the Queen says consolingly: 'You see, I've been doing it for forty years.'

Crying, no known reason. The pillow wet with the tears of things.

Waiting for the end of the world as announced. Touching adieus in the darkness. Everyone sad but brave. A great dreadful flash of light. Silence. Everyone is still alive.

Lots of old people slumped in chairs. It must be an old people's home. What are you doing there? You don't feel at home.

You are on your deathbed. You console relatives and friends. You utter noble sentiments. It is all richly satisfying. But then it strikes you that you don't in the least want to die.

You are rubbing your eyes, as if you can't believe them. The next day you are reading a famous German, who says that life and dreams are pages in one and the same book and in sleep we merely read haphazardly. You find you are rubbing your eyes.

Warnings, warnings!

Many of you will not have purchased shares before. You would never have dreamt of it. Shares were not for the likes of you. Now you can buy shares as you like, and we hope you will. But bear in mind that SHARES CAN GO DOWN IN VALUE AS WELL AS UP. For PAST PERFORMANCE IS NOT NECESSARILY A GUIDE TO THE FUTURE, and it is our duty to warn you that shares may disturb the balance.

*

Many of you will be having sex for the first time. You would never have dreamt of having it before. Sex was not for the likes of you. Now you can have sex as much as you like, and we hope you will like it. But bear in mind: SEX CAN GO DOWN IN VALUE AS WELL AS UP, and PAST PERFORMANCE IS NOT NECESSARILY A GUIDE TO THE FUTURE. We must warn you that sex can affect the heart.

*

Many of you will never have drunk champagne before. Please note that the liquid in the bottle is under severe pressure. CHAMPAGNE CAN GO UP AS WELL AS DOWN. In no circumstances should the cork be removed with a corkscrew. You are advised to take it between thumb and forefinger and ease it out gently. Champagne corks may have far-reaching consequences.

*

Many of you will not have lived before. You would never have dreamt of it. Now you are living, and we hope you will enjoy it. But it is our responsibility to warn you that LIFE CAN GO DOWN IN VALUE AS WELL AS UP, and the past is not a sure guide to the future. Living can be bad for your well-being and even lead to death.

The great man can be quite funny, nicht wahr?

1

For example, a scene in Heaven, which opens
With the Heavenly Hosts singing their hearts out.
I can't sing, I'm down to earth, in fact well up in it.
At last His Lordship spies me and raps the music stand.
Do I know a certain professor? I certainly do,
Him and his seminars, teaching young ideas to shoot.
While a man still tries, you say, still must he stray.
That one's decidedly trying. Ought to stray a treat.
You can stomach irony? And you don't really mind
If I'm a bit of a devil—devils are creative?
I'll create, I'll be your left-hand man. Adieu for now . . .
Sadly out of touch, the old toff, but civil enough,
It's not every god who'll chat with the likes of me.
Must arrange for a garden, like the one my forebear
Found lovesome, basking in the grass, loafing under trees.
Then, one fine evenfall, a diffident daughter of Eve,
And the professor himself, after wetting his whistle
In the SCR. Good for a laugh, you have to admit.

2

Need to be asked in on the first occasion. It's a rule. Hence assume shape of a publisher's scout ('You being the leading expert on Frauenlob . . .'), or a door-to-door apothecary ('These are uppers, those are downers, I can recommend both'), or a damsel in distress ('My carriage has lost a wheel, may I prevail upon . . . ?'), or, best of all, a dog, a small friendly one ('Yap, yap, yap'). Dogs are fun to do, and people round here love them. Well worth the trouble. Two souls dwell in his breast, he says. Two for the price of one.

3

It shouldn't be difficult to spot the humour.
To wit, a don in a garden with a pretty wench:
'Don't let's talk of intellectual matters.'
He can do that elsewhere, innocence is more exciting.
So her brother's a soldier? But away on manoeuvres.
She's demolishing a daisy: 'He loves me, loves me not . . .'
(The petals work out right, the Other saw to that.)
She's on her feet all day? 'Lie down awhile, my dear.'
Does he believe in God? She's so sweet, so artless,
Oh lordy! (Nor's the Lord in this neck of the woods.)
It's time for a little speech, something irenic
(The ironic he'll leave to the Other). 'Who can say . . . ?
What do we mean when . . . ? Names were deceivers
 ever . . .
We each have our way . . . Plurality it's called.'
(It's called blinding with scholastic insights.)
'God is love. Love is god.' And oh yes, 'Feeling is all!'
He feels. She is felt. And that was the garden scene.
Perhaps it's not too hard to miss the comic side.

'You're mine thereafter'

'Not twenty-four, but fourteen years . . .
It's tempting still—so what's the price?'

'You've read the literature,
You know the going rate.'

'Something valueless or else invaluable,
Either whatever never was, or else
What always will be.'

'Let's not talk of market forces.
As prices go,
The price is modest.'

'To disagree would be immodest in me.
Yet if you really want it
The commodity must have its worth.'

'Call it a whim of mine,
Call it collector's mania,
Call me a rag-and-bone man—
White-collar branch.'

'Fourteen years? When all the drink in Christendom
Can't make me think that every girl's a Helen—
Or that we're speaking verse? I'll leave it,
If you'll forgive the saying, in God's hands.'

'You think his hands are kindlier than mine?
Ask yourself: what have you got to lose?
The chances are that in the end I'd get you
Gratis, a godsend, no strings attached.'

'Again I hardly dare to contradict.
Unreasonably, hearts have their reasons.
Call it a whim of mine,
Call me a body-and-soul man.'

'That's fine by me. I'll call back later,
When time has grown more precious to you.
Though fourteen will have dropped to four,
I fear, or years to months . . .'

'And months to days . . . How fast they all go by,
Days, months, and years! Make me immortal
—Ah, you can't—or leave me as I am.'

As if human

It was as if
It was as if you were sensible
As if you were insensible
It was as if you loved
Or at least you liked
(As if you could tell the difference)
It was as if you took pleasure in
As if you took umbrage at
It was as if you were preaching
As if you were practising
It was as if you slept on roses
As if you slept on thorns
As if (or was it) you were dreaming
As if you woke up in the morning
As if (or if not) you remembered
It was as if the cap fitted
As if you didn't wear a cap
As if you could keep your head
As if you didn't have one
It was as if you were virtually real
As if you were really virtual
It was as if you led a double life
As if you led a single one
Or was it as if you were dead
(It wasn't as if you rested in peace)
It was almost as if
Ifs and an's were pots and pans
Almost as if as if were is
(Or was it)

Bindings

Books came to pieces in the tropics.
Which of them returned the stouter?—
The ones patched up in Changi Prison
(Like Eliot, Pope and Chaucer),
Or those on less demanding topics
(Margins guillotined to match the cover)
Trusted to the Psychiatric Institution?
All, I think, will last for ever.

A register of diverse deaths

Her sons having gathered at the bedside,
'I am dying, as you see,' she informed them.
It was something they could hardly dispute.
'I am buried in my body': the which
Had grown as refractory as a coffin.
'And so,' she added, 'I wish to be cremated.'
An option the sons didn't entirely approve
In view of the resurrection of the body,
In which she was already as good as buried.
'But don't attempt to scatter the ashes—
They'd only blow back on you. Brick them up.'
The mischievous spirit paying off old scores?
Or a mother concerned for neat appearances,
Or simply to assure them that she wouldn't cling?
'Nothing undisclosed remaining,' the Mass would say,
She fell asleep, they adjourned the meeting *sine die*.

A book at bedtime

Thinking of the ten-year-old, hydrocephalic,
Who longed to finish *Pickwick* before he died
(And prayed to God to let him, and got halfway through),

He decides to embark on *The Posthumous Papers*,
Prudently reading backwards, last chapter first
(But without seeking God's prior agreement).

On second thoughts he reckons it had better be
A short story. 'The Ghost's Bargain'? No, make it
Great Expectations, backwards, beginning at the end.

So many

Caught from the corner of the eye,
Sketching themselves in pipe smoke,
Seen in dreams, or glimpsed awake—
Death, you knew, had undone so many.

What we never and always think of

And by no means wrongly in either respect,
Since it is only right or vital we should fight it;
In such strife we find life's larger portion.
It lies in our being never to acquiesce;
Less so, though at times we can, to accept not being
(Paradox from the start was a peculiar part of us).
And in the end that we are destined to lose
Is no fearful matter, if naturally a matter
Of fear, fear of the unknown which we constantly seek
To know. There seems no distinctly superior outcome:
Immortality, despite our dreams, is eternal dying
(Other things alas are also never-ending,
The everlasting asks for inconceivable change);
Better to cease to live while still living.
Think of yourself as a species that wasn't quite fit.
Mother Nature knows best. She is busy, we must hope
She is minded to be reasonably kind.
It was mothers who gave us birth? But sophistry,
However pleasing, is only a tease. And the topic,
The poet reckoned, is one we ought to be ready
To be blank about—ah, but we aren't!
We're not ready for anything, nor for nothing.
We have read the books, and the soul remains doleful,
Whatever the sum of years awarded by fluke or favour.
Lyrics have had their day, or should have, and this
Is scarcely stuff to lyricize on. We 'stoop to truth':
So rather an issue to deal with (brave words!) in prose,
A sentence, lengthy or short, that one day stops dead,
Its rhyme or chime right out of earshot, or not in sight
Yet, or . . .